Know
Your
Chickens

Jack
Byard

Publishing

First published 2010,reprinted 2010, 2012

ISBN 978-1-906853-30-3

Published by
Old Pond Publishing Ltd
Dencora Business Centre
36 White House Road
Ipswich IP1 5LT
United Kingdom

www.oldpond.com

Book design by Liz Whatling
Printed and bound in India

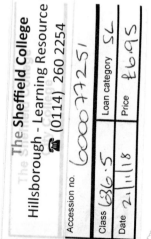

Contents

Acknowledgements

Thank you to all the poultry breeders and enthusiasts who have offered or supplied photographs for use in this book. A special thank you to David Brandreth and the Devonshire Traditional Breed Centre whose kind offers of photographs reduced my stress level and telephone bill by half, making my wife, Elaine, eternally grateful. Thank you also to my 11-year-old granddaughter, Rebecca, whose input keeps me grounded and focused on the younger readers. Once again, any mistakes are mine and mine alone.

Picture Credits

Plate *(1)* Philip Smedley, *(2)* David Brandreth, *(3)* Big W Ranch, *(4)* David Brandreth, *(5)* W&K Pimlott, *(6)* David Brandreth, *(7)* David Brandreth, *(8)* W&K Pimlott, *(9)* David Brandreth, *(10)* David Brandreth, *(11)* Pembroke Plants and Poultry, *(12)* David Brandreth, *(13)* The Chicken Garden, *(14)* David Brandreth, *(15)* Big W Ranch, *(16)* Big W Ranch, *(17)* John and Sylvia Cook, *(18)* South Yeo Farm East, *(19)* David Brandreth, *(20)* Robert and Veronica Potter, *(21)* Maria Castro, *(22)* David Brandreth, *(23)* David Brandreth, *(24)* David Brandreth, *(25)* David Brandreth, *(26)* Brockscombe Valley Farm, *(27)* Mr Benjamin Crosby, Britannic Rare Breeds, *(28)* David Brandreth, *(29)* Rising Sun Farm, *(30)* David Brandreth, *(31)* David Brandreth, *(32)* Rodriquez Poultry, *(33)* David Brandreth, *(34)* David Brandreth, *(35)* David Brandreth, *(36)* Brockscombe Valley Farm, *(37)* Brockscombe Valley Farm, *(38)* Elaine Hargreaves, *(39)* David Brandreth, *(40)* Wanda Zwart, *(41)* David Brandreth, *(42)* South Yeo Farm East, *(43)* David Brandreth, *(44)* David Brandreth, *(title page)* Wanda Zwart.

Foreword

Our feathered companions, chickens, have been around, as an old farming friend of my father would say, 'since Adam was a lad'.

Their rich diversity of colour, size and feather pattern and their elegant, toy-soldier-like strutting are a joy to behold. Most are friendly, docile creatures but there are, as amongst us, a few grumpy old men who will look at your legs as the next hors d'oeuvre. Many are bred for the sheer pleasure they give to the breeders who will, after a full day's work, put in another 2 or 3 hours taking care of their charges.

It is commonly believed that Britain's commercial farms and farmers are the best in Europe for animal welfare and ensuring that

all animals have a happy and stress-free life. If, like me, you consider animal welfare a priority, please buy British whenever possible.

Some of the chickens in this book are rare and I have used the Rare Breeds Survival Trust gradings to show this. The gradings only apply to the colour of chicken pictured – a different colour of the same breed may not be as rare.

JACK BYARD
2010

Ancona

Native to
The Mediterranean

Now Found
Throughout the British
Isles, Europe and America

Protection category

Description

The Ancona is one of the oldest breeds of chicken and arrived in the British Isles in the mid to late 19th century from Ancona in Italy. The breed bears a strong resemblance to the Leghorn so was once also known as the Black Leghorn or, because of its patterned feathers, the Mottled Leghorn.

The Ancona is extremely popular in Europe. These tough, hardy birds can adapt easily to a range of environments and are great scavengers with an of instinct for finding food. They are also well known for their egg-laying abilities since it is quite common for an Ancona to lay 300 white eggs a year. These many skills mean the breed has flourished.

The Ancona must always be kept free range but surrounded with a high fence or this high-flying bird will soon disappear. They are no longer used commercially but still have a firm place in a domestic flock and at poultry shows.

Colour

The plumage is mottled black with white-tipped feathers and often has a lustrous green tint. More white appears as they get older; sounds familiar. The face, comb (which can be single or rose) and wattles are red. The earlobes are white and the beak is yellow with black and fawn markings. The legs and four-toed feet are yellow with black mottling. The eyes are orange-red with bay pupils.

2.

Andalusian

Native to
Andalusia in Spain

Now Found
Throughout the British Isles, Europe, North America, Canada and Australia

Protection category

Description

This ornamental bird began in Andalusia but the breed was further developed in the British Isles and North America. The modern blue Andalusian is a result of crossing black and white birds imported from Andalusia in 1846. Crossing two blue birds will result in around 25% black offspring, 25% white and the others will be blue. Because of this low incidence of the desired colour, the Andalusian is only bred by enthusiasts with an interest in preserving the breed. The Andalusian lays in the region of 160 creamy white eggs a year.

The Andalusian has a magnificent presence. It is elegant and graceful with a carriage of which any catwalk model would be proud. They are extremely fast runners; the breed society suggests you invest in a landing net. When people are asked why they keep Andalusians the answer is nearly always 'because they are elegant and beautiful'.

Colour

The ideal colour is slate blue with black lacing on each feather. Both males and females have black hackles. The legs, feet and four toes are slate blue or black. The beak is slate or fawn coloured. The eyes are red or reddish brown; the earlobes are white and the wattles, face and large single comb are red. The hen's comb flops to one side.

Appenzeller
Spitzhauben

Native to
Switzerland

Now Found
Throughout the British Isles, Europe, America, Canada and Australia

Description

The Appenzeller originated in the Swiss Canton of Appenzeller and is thought to have existed for over 400 years. It is the national breed of Switzerland. The Spitzhauben variety has a pointed comb similar to the traditional pointed lace bonnet of the area also called a Spitzhauben.

During World War II, the Appenzeller came close to extinction and it was only the dedication of German breeders in the 1950s that ensured their survival.

This is a good, hardy breed which is well adapted to living in mountainous regions and requires little attention. If they have a good foraging area they will almost look after themselves. Appenzellers do not suit confinement and are at their best when given the freedom to roam. They are superb climbers and will happily roost in trees.

They lay in the region of 150 white eggs per year and are considered to make good companions.

Colour

The Silver Spangled Appenzeller has a background colour of silvery white with black spangles (irregular shaped dots) on the feather tips. There are fewer dots on the neck and head. The horn-type comb and wattles are red, the beak is black or fawn and the earlobes are white. The eyes are brown and the legs are blue or black. The Appenzeller also has a Barthuhner variety. There are three colours of each: Gold Spangled, Silver Spangled and Black Spangled.

Araucana
Lavender

Native to
Northern Chile

Now Found
Worldwide

Description

The Araucana is an ancient breed which was named after the Araucan Indians who lived on the plains of the Andes Mountains in Chile.

In the early 16th century, the Portuguese explorer Magellan recorded poultry resembling the Araucana. Later in the same century the breed arrived in countries surrounding the Mediterranean. The true-breeding Lavender Araucana was developed in Scotland in the 1930s by George Malcolm.

The Araucana does not lay many eggs but the eggs they do lay are spectacular in colour. They are mainly blue or green but can range from greenish blue to violet blue to greyish. The shell colour is unique in that it is the same colour inside and out. The Araucana are hardy, grow quickly and mature fast. They are content to be in a pen as long as there is a regular supply of fresh grass.

It is believed that in 2006 the Spangled variety became extinct. I hope this proves to be wrong and in some corner of the world there is a Spangled Araucana.

Colour

They are blue-grey with a small pea comb and face which are bright red. The eyes are dark orange; the beak and claws are creamy fawn and the legs can be slate, olive or willow. There are two varieties: Rumpless - without a tail but with feathered tufts growing out from near the ears and Tailed - with feathered muffs. The other colours are: Blue; Black/Red; Silver Duckwing; Golden Duckwing; Blue/Red; Spangled; Cuckoo; Black and White.

Australorp
Black

Native to
Australia

Now Found
Throughout the British Isles
and on most continents

Description

The Australorp was developed using the Black Orpington which was imported into Australia from the British Isles in the late 19th and early 20th centuries. The Australian breeders wanted a quality dual-purpose breed and this was achieved by crossing the Black Orpington with the Langshan, Minorca and White Leghorn. The Australorp was born.

Several people wanted the honour of choosing the name since Australian Laying Orpington did not trip lightly off the tongue. In 1919 Arthur Harwood suggested Austral with an added 'orp' to honour the ancestors. In 1921 the breed was imported into the British Isles.

This is a good all-round breed, laying in the region of 250 light brown eggs a year. A group of six Australorps hold the world egg-laying record of 1857 eggs (an average of 309.5 eggs per hen) over 365 consecutive days in 1922 and 1923.

They are docile, friendly and happy in a run as well as free range. They make a good choice for a beginner.

Colour

Black but in the sunlight its feathers gleam green and purple. The face, comb, wattles and earlobes are red. The eyes, feet and legs are black, except the soles of the feet which are white. The other colours are Blue and White.

Barnevelder
Double Laced

Native to
Holland

Now Found
Worldwide

Description

Developed during the 19th century in Barneveld, the breed was originally created to produce beautiful dark brown eggs.

The breeds used to create the Barnevelder were chosen with great care. A Dutch hen was crossed with a Langshan which produced the dark eggs. Further crosses with the Buff Orpington, the Brahma and the Cochin were purely to improve the quality and colour of the eggs. The bird's feather pattern was immaterial at that time but from 1920 attempts were made to standardise the breed. In 1921 the breed was imported into the British Isles and the Society of Barnevelder Breeders was formed.

The growing fame of the breed led to worldwide exports. This hardy, dual-purpose hen is capable of laying an average of 200 dark brown, good-flavoured eggs each year. The Barnevelder is a lazy chicken and must be kept free range to ensure it receives plenty of exercise. Overweight Barnvelders do not lay eggs.

Colour

The Barnevelder has double-laced mahogany and black feathers. The male's black neck feathers glow with shades of green and have a red-brown edging. The main body colour is reddish brown. The comb, face, wattles and earlobes are red. The prominent eyes are orange and the legs and four-toed feet are bright yellow. The beak is yellow with a black dot except on the Silver variety which has a fawn beak. The other colours are Black, Partridge and Silver.

Belgian d'Uccle
Mille Fleur

Native to
Belgium

Now Found
Throughout the British Isles, Europe, Australia and America.

Description

The Belgian d'Uccle (pronounced 'dookluh') was developed in Uccle on the south-eastern border of Belgium by Michel Van Gelder in the late 19th century. The Mille Fleur (meaning 'A thousand flowers') was one of the first varieties of the d'Uccle breed. It is understood that the breed is a cross between the Dutch Sabelpoot Bantam and the Antwerp Bearded Bantam. However, Van Gelder was a frequent visitor to British and German poultry shows so the possibility that breeds from these countries influenced the d'Uccle cannot be ruled out.

To the uninitiated, myself and even some poultry breeders included, the d'Uccle and the Booted Bantam are the same. Wrong. The Booted Bantam has no beard and large red wattles while the d'Uccle has a beard, muffs and small or non-existent wattles. As the meerkat says, 'simples'.

A docile and friendly breed. To quote one breeder, 'I found these birds to be so tame it was often a problem to stop them flying upon me when I did not want to be seen wearing chickens.'

Colour

The Mille Fleur has a light mahogany background colour. Each feather has a black spot and is tipped with a V-shaped white spangle. The combs and earlobes are red and the eyes are orange-red with black pupils. The legs are feathered and the skin and feet are blue-grey. Other colours are: Black; Blue; Lavender Quail; Silver Quail; Blue Quail; Cuckoo; Porcelain; Lavender Mottled and Black Mottled.

Black Rock

Native to
America

Now Found
Throughout Britain
and on most continents

Description

The Black Rock is a hybrid, the result of a Rhode Island Red male and a Plymouth Rock female. The gold-to-ginger feathers on the chest and underneath vary in intensity from bird to bird and are inherited from the Rhode Island Red. Occasional silver colouring comes from the Plymouth Rock. The sex of the Black Rock can be determined at a day old and as a result the male Black Rock is rarely seen.

This attractive, hardy breed, with its thick plumage, is more than capable of coping with the vagaries of the British weather. The breed has a naturally highly developed immune system.

This busy, inquisitive breed will roost in a tree if given half a chance. It is easy to handle and an ideal pet for older children. They will lay up to 300 medium to dark brown eggs a year. The Black Rock enjoys freedom and becomes bored if confined.

Colour

The body is black with a green sheen. There is gold feathering on the hackle and the chest has red and gold feathers tipped with black. The face, wattles and earlobes are red and the small semi-erect comb is red with black spots. The eyes are brown, the beak is black and the legs and feet are slate grey.

Brahma
Gold

Native to
America

Now Found
Throughout the British Isles
and on most continents

Description

The Brahma is named after the Brahmaputra River in India
but for many years its true origin has been in doubt. Most
breeders now agree that the breed developed in America
and that the Shanghai, imported from China in the 1840s
and now known as the Cochin, was used to improve the
Indian Grey Chittagong which was then renamed the Brahma.
A group of nine birds arrived in the British Isles in the mid
19th century as a gift for Queen Victoria. Their appearance
caused a sensation. In 1853 a Dark Brahma was sold for £105,
a huge sum in the days when an ordinary agricultural
labourer earned about 9s. 3d (46p) a day.

The Brahma, a dual-purpose breed, will lay in the region of
150 brown tinted eggs a year. Today they are mainly bred for
ornamental use. The feathered legs and feet must be kept
dry; they collect mud and if this is allowed to dry it can cause
serious damage to the feet and legs. Described as large,
stately, docile and trusting, they make great pets which will
roam happily in a garden with a low fence.

Colour

The head is a rich gold which continues down the neck. Each feather has a central black stripe, the body is a dull red-black and the tail a shiny black. The back is gold coloured, the under-body is glossy black and the wing bows are bright red. The face, comb, earlobes and wattles are red and the eyes are orange-red. The feet and legs are covered in feathers and bright yellow underneath. Other colours are: Dark; Light; White; Blue Partridge and Buff Columbian.

Cochin
Buff

Native to
China

Now Found
Throughout the British Isles
and on most continents

Description

The Cochin, or Shanghai as it was originally known,
originated in China over 150 years ago and was imported
into America and the British Isles in the 1800s. The first
arrivals were presented to Queen Victoria and the
appearance of these large, friendly 'balls of fluff and
feather' created quite a stir. It bore no resemblance to
any known breed at that time. The Cochin became an
overnight success and a must-have. According to *Punch*
magazine, in 1853 one sold for £2,587. It was known as
the Cochin Craze.

Further Cochin were imported and the breed was
developed, continuing the work that had already been
carried out in America.

The Cochin is hardy to the cold but in summer its thick
feather coat can cause over-heating. A gentle spray with
water will keep them cool, happy and healthy. They will
lay around 100 eggs per year.

Colour

The male's head, hackle, back, shoulders, wings and tail are blending shades of cinnamon, gold and lemon. The face, comb, wattles and earlobes are red and the eyes match the feather colouring. The legs and feet are bright yellow. The hens are a richer colour. Other Cochin colours are: Black; Blue; Cuckoo; Partridge and Grouse, and White.

Croad Langshan
Black

Native to
Asia

Now Found
Throughout the British Isles, Europe, America and Australia

Protection category

Description

The Croad Langshan is a development of an ancient breed from the Langshan district of Northern China where the original breed can still be found. The Maran, the Barnevelder and the Black Orpington were all developed using the Langshan. The Langshan was imported into the British Isles in 1872 by Major F T Croad. The Croad was added to the name as a tribute to the Major's niece, Miss A C Croad, who gave unstinting support during the development and improvement of the Croad Langshan. The breed club was opened in 1904.

In the early 1900s the Croad Langshan was a popular dual-purpose breed. After World War II, as with many other breeds, the numbers went into serious decline and it was only the intervention of the Rare Poultry Society that prevented extinction.

The breed prefers dry, sheltered conditions and adapts to both a free range life and to smaller open enclosures. The Croad Langshan lays up to 200 eggs a year which are brown with a plum-coloured bloom. Graceful, intelligent and docile. No, not me, the beautiful, inquisitive Croad Langshan.

Colour

Black with a beetle-green lustre. The single comb, face, wattles and earlobes are red and the beak is light to dark fawn with grey streaks. The legs and toes are blue-black. The legs and outer toes are lightly feathered. Other colours are White and Blue.

Dorking
Silver Grey

Native to
Britain

Now Found
Throughout the British Isles, Europe, America and Australia

Protection category

Description

Bred in Italy during the reign of Julius Caesar (100 to 44BC), the Dorking is one of the oldest breeds of domesticated poultry. They have five toes, whereas most breeds have four. This superb dual-purpose breed was brought to the British Isles by the Romans and it is here that most of the development and improvement has taken place. It made its debut in 1845 at a British poultry show - well, these things take time!

The Dorking has been used to develop many modern breeds. It is one of the few breeds with red earlobes to lay white eggs and it lays in the region of 140 a year. The Dorking is a large, docile bird. In order to grow and remain healthy it must have a large area in which to exercise and forage. It produces excellent meat and eggs.

When you look at the Dorking, you are looking at history but the breed is rare with fewer than 500 breeding pairs known to exist.

Colour

The male is outstanding in appearance. He has a silver-white hackle and saddle with black on the lower body. The hens are gentle shades of slate grey with a black-striped silver hackle and a salmon breast. The beak is white or fawn. The legs and five-toed feet are white with a pinkish tinge. Other colours are: Red; Dark; Cuckoo and White. All variations have red eyes, wattles, combs and earlobes.

Dutch Bantam
Gold Partridge

Native to
The Netherlands

Now Found
Worldwide

Description

The Dutch Bantam is a true bantam which means it has no large fowl equivalent.

The breed has been around for hundreds of years. Old Dutch painting show chickens similar to the Dutch Bantam, as do British paintings of the 1860s.

Writings indicate that it was introduced to Holland by sailors who collected them from the Bantam Islands in the Dutch East Indies in the 17th century.

The popularity of the breed in its early years is attributed to the law that all large eggs had to be given to the Lord of the Manor. The small eggs of the bantam escaped this demand.

The Dutch Bantam is active and, if handled regularly, becomes very friendly. Because of their diminutive size they are ideal if you do not have golden acres. A good aviary or a high fence is a necessity as they are good flyers.

Colour

The male's hackle shades from dark to light orange and each feather has a dark green central stripe. The breast is black with a green sheen, the shoulders are a deep red-brown and the wing bar is iridescent green. The main tail feathers are iridescent green-black. The face, single comb and wattles are red, the earlobes are white, and the beak is dark fawn or has a blue tinge. The legs and feet are slate grey. Other colours include: Silver Partridge; Yellow Partridge; Blue Silver Partridge and at least another twelve colours.

Faverolles
Salmon

Native to
France

Now Found
Throughout the British Isles
and on major continents

Description

The Faverolles, from the village of Faverolles in Northern France, were developed from a cross between the Houdan, the Cochin and the Dorking. This ancestry can be traced by the fact they have five toes. There are three different types of Faverolles, the original French, the German and the British. They were initially imported into the British Isles in 1886 and were subsequently crossed with the Orpington, the Sussex and the Indian Game to produce the dual-purpose breed we know today. A French writer said, 'as farmyard fowls they stand unrivalled, their superiority being uncontestable'.

This genteel, sweet-natured bird is an ideal breed for children and has been described as the Peacock or French Poodle of the chicken world. Whatever its description, it certainly stands out in a crowd with its creamy white beard and muffs.

They lay in the region of 160 light brown or creamy eggs throughout the year.

Colour

The female is a mixture of brown and creamy-white, known as Salmon. The other colours are: Black; Blue Laced; Buff; Cuckoo; Ermine and White. The males are iridescent black with a bronze back, black wings and straw coloured hackles. The faces, combs and wattles are red and the eyes are orange to yellow. They have white feathered legs and five-toes.

Golden Phoenix

Native to
Japan

Now Found
Worldwide

Description

This striking breed originated in Japan where it was known as the Onagadori. It was one of several long-tailed breeds that have been bred in Japan for a thousand years and kept in the Imperial Gardens. The tail feathers of this exceptional breed could reach 10 metres in length.

In the 1800s the Golden Phoenix was crossed with German game birds in order to improve the health and wellbeing of the breed. Further development and improvement was carried out in America using the Leghorn. This crossing had the effect of breeding out the gene responsible for the exceptional tail feathers but a modern mature male can still have a tail up to a metre long.

The Golden Phoenix is well adapted to free-range life but this can be at the cost of the tail feathers. A good roost well clear of the ground will give them the best of both worlds.

Golden Phoenix hens are good mothers, docile but not too friendly, and have been known to chase off marauding cats taking too close an interest in their brood.

Colour

The amber of the male's head continues down the neck, shading to a golden buff at the shoulder. The wings and chest are green with downy white fluff on the saddle just in front of the tail. The tail feathers are black but glow purple-green in sunlight. The face, single comb, wattles and earlobes are red, the beak is a light tan and the legs and feet are slate blue.

Hamburgh
Silver Spangled

Native to
Holland and Germany

Now Found
Throughout the British Isles, Europe, America and Australia

Protection category

Description

Where, oh where do you come from? With a history dating back to the 1600s it is not surprising that the country of origin, despite the name, is unclear. The spangled Hamburgh was developed in Yorkshire and Lancashire in the 1700s.

This small breed is capable of laying around 150 eggs per year. It was once known as the Dutch Everyday Layer but its use is now mainly ornamental. They also go by the name of Moonies because of their spangles.

Stylish, elegant and snappy are adjectives often used to describe this beautiful bird. Its appearance will brighten even the darkest of days. The Hamburgh thrives best when free range so please don't confine them. They are good foragers and capable of flying a good distance.

Colour

The plumage has lustrous greenish black spangles on a silvery white background. The end of each tail feather has a half-moon shaped black spangle and half way up the wing feathers are round black spangles forming a row across the wing. The rose comb, face, eyes and wattles are red and the ear lobes are white. The legs and feet are leaden blue. Other colours are: Black; Gold Pencilled; Silver Pencilled and Gold Spangled.

Indian Game
Dark

Native to
Cornwall

Now Found
Worldwide

Protection category

Description

The Indian Game is a beautiful old British breed. It was developed in Cornwall in the early 19th century by Sir Walter Raleigh Gilbert who crossed the ancient fighting breeds of Red Asil, Malay and Old English Game to produce fighting birds. It turned out to be too heavy for a good fighter but, in any case, cock fighting was banned in England soon after in 1835. The breed was first shown at the Crystal Palace in 1858 but was little seen outside Cornwall until the end of the 19th century.

Its true worth was realised in its excellent meat, a trait that is valued to this day. It is crossed with the Dorking and Sussex breeds with superb results. Although poor layers, producing only about 80 light brown eggs per year, they are attentive mothers. Today, in its purest form, this proud, elegant bird can be seen at poultry shows worldwide.

Colour

The females are a rich mahogany colour with glossy black double lacing around each feather. The males are glossy black with a trace of deep chestnut in the wings. The beak is yellow or fawn striped with yellow. The face, pea comb, wattles and earlobes are neat and red. The legs are yellow or orange and the eyes pale yellow to pale red. The other colours are Jubilee and Blue Laced.

Ixworth

Native to
Suffolk

Now Found
Throughout the British Isles

Protection category

Description

Developed in the Suffolk village of Ixworth, hence the name, by Reginald Appleyard. He began his work in 1930 and the Ixworth breed was finally unveiled in 1939.

The Ixworth is a cross of several breeds: the White Old English Game; the Jubilee Indian Game; the White Sussex, the Orpington and the White Minorca. The result is a fine bird with a white skin ideally suited to the requirements of the British market. The Ixworth also has a good egg-laying record. This is a good, all-round breed that deserves much greater recognition.

The Ixworth came within a hair's breadth of extinction between 1950 and 1970 and it was left to a few hardy breeders, mainly in the Shropshire area, to keep the breed alive. It is still not out of danger but numbers are increasing. Let us hope that more people take to breeding this excellent bird. The bantam version is now extinct.

Colour

Always white with pinkish white legs, feet and beak. The eyes are orange to red. The pea-type comb, face, wattles and earlobes are red.

Japanese Bantam
Black Tailed White

Native to
Japan

Now Found
Throughout the British Isles
and on most continents

Description

The Japanese is a true bantam, having no large fowl
equivalent. The development of the Japanese Bantam or
Chabo (a Javanese word for dwarf), dates back to at least
the 7th century. It is a cross of breeds from China and South
East Asia and could once be seen around the gardens of
rich and famous Japanese. The breed arrived in Europe in
the 1700s and was a feature in both Japanese and Dutch
art in that period.

The Japanese Bantam has very short legs and when walking
appears duck-like because of its U-shaped profile and
rounded shape. This small, attractive bird has a large tail
which reaches above its head by a third of its height. The
Japanese Bantam is an ideal garden chicken because its legs
are too short to do too much damage to your lawn. They
are calm, docile and friendly and so make an ideal pet for
children. With good care they will live over ten years.

Colour

They are pure white with a red face, single comb, wattles, earlobes and eyes. The legs are yellow and the tail is black. The other colours include: Black Tailed White; Black; Blue; Brown Red; Grey (Birchen, Silver, Dark and Millers); Mottled (Black, Blue and Red); Silver Duckwing; Gold Duckwing; Wheaten; Black Breasted Red; Blue; Lavender; Cuckoo; Red and Tri-coloured. I understand there are thirty-five colours in total.

Jersey Giant
Black

Native to
New Jersey, in America

Now Found
Throughout the British Isles
and on most continents

Description

The breed was developed in the late 19th century by the brothers John and Thomas Black of Burlington County in New Jersey and accepted as a breed in 1922.

The breed was developed by crosses of the Dark Brahma, the Black Java, the Black Langshan and possibly the Black Orpington. The Black variety was followed in 1947 by the White and it was not until almost 40 years later that the Blue variety appeared.

This is a large bird, originally bred as a table bird that would be an alternative to turkey. A mature Jersey Giant weighs in at an average of 5.90kg. It is a superb dual-purpose breed; it lays in the region of 160 large light to dark brown eggs a year which are in keeping with the bird's size.

Jersey Giants are cold-hardy and, despite their size, calm and gentle. They are too slow growing for commercial breeders but quality is worth waiting for. The world's largest chicken breed.

Colour

The body is shiny black with a green sheen and the under-body is slate grey. The face, single comb, wattles and earlobes are red. The beak is black with a yellow tip and the eyes are dark brown. The legs and feet are black turning to a dark yellowish green in mature birds. The other colour variations are White and Blue.

Lakenvelder

Native to
The Netherlands

Now Found
Throughout the British Isles
and on most continents

Description

The Dutch believe the breed is named after the village of Lakenvelt. They point out that they also have breeds of goat and cattle with black heads and tails called Lakenvelders, which translates as 'shadow on a sheet'. The Germans argue that it may have a Dutch name but it was developed in the Westfalen area of Germany. Records show that the Lakenvelder was around in the early 18th century but did not appear in the British Isles until the beginning of the 20th century.

One breeder describes them as being 'independently minded little souls, who prefer to forage rather than eat the grain supplied and will roost in trees in all weathers.' The Lakenvelder is a fairly small bird - the male weighs about 2.5kg - but very active and adaptable. They will live contentedly in a confined area but are much happier free range where they can forage. They lay an average of 160 white or creamy white eggs per year.

Colour

They have a black head, neck and hackle. The remainder of the body is mostly white with a slate grey under-colour. The face, comb and wattles are red, the earlobes are white and the eyes are red or orange. The beak is dark fawn. The legs and feet are slate grey. There is also a blue variety known as Blue Marked.

Legbar
Cream

Native to
The British Isles

Now Found
Throughout the British Isles, Australia and New Zealand

Protection category

Description

The beginnings of the Legbar date to the early 20th century. Clarence Elliott, botanist and explorer from Stow-on-the-Wold in Gloucestershire, returned from Patagonia with three hens and a cockerel of an unknown breed. The three hens survived but as a result of a breakdown in translation the old fellow was cooked for lunch.

In 1930 the three hens were in residence at Cambridge University under the care of Professor Reginald C Punnett, an expert in poultry genetics. The Patagonian trio was crossed with the Brown Leghorn, the Barred Rock and the Araucana and, after some further development, they became the modern Legbar.

The Legbar is known as an autosex breed - the colour of the fluff soon after hatching indicates the sex. They lay in the region of 180 sky-blue or olive-green eggs a year but, unlike its Araucana ancestor, the shell colour is on the outside only. They are vigorous, sprightly, alert and happy in small or large runs; if kept in a large run, trying to catch one will reduce your need for the gym.

Colour

The male has a cream hackle with a small amount of grey barring. His back is cream with grey and occasionally chestnut bars. The wings are a creamy grey with chestnut bands and the large tail is grey and white with chestnut bars. The face and wattles are red, as is the comb which is large and floppy with a tuft of grey feathers behind it. The earlobes are white to cream and the feet, beaks and legs are yellow. The other colours are Silver and Gold.

Leghorn
Brown

Native to
Italy

Now Found
Worldwide

Description

The Leghorn is named after the Italian city of Livorno. It found its way to America in the 1830s and arrived in the British Isles at the latter end of the 19th century. It was here that most of the colours were developed. The Leghorn was then crossed with the Malay and the Minorcan to improve the quality and size of the eggs.

The Leghorn is well represented in the world of advertising. Most cockerels depicted are slim with large combs and long curved tail feathers. Foghorn Leghorn would be proud.

The breed is one of the best known breeds and produces the majority of the world's supply of white eggs. A good White Leghorn will lay in the region of 300 eggs a year, the Black and Brown laying slightly less, about 200 to 250 a year. A strutting, noisy bird with presence. This good forager is capable of flying a good distance and will happily roost in a tree.

Colour

The head and hackle are orange, shading down to pale yellow. The back, shoulders and wings are deep red or maroon. The breast and the remainder of the body are glossy black and the tail is black and shimmering green. The face, eyes and comb, which is single or rose type, are red. The earlobes are white and the legs are yellow. Other colours are: Black; Buff; Cuckoo; Blue; Golden Duckwing; Silver Duckwing; Exchequer; Black Mottled; Red Mottled; Partridge and, last but not least, White.

24

Lincolnshire Buff

Native to
Lincolnshire

Now Found
Throughout the British Isles

Description

The Lincolnshire Buff was developed in Lincolnshire over 150 years ago by crossing the Cochin and the Brahma with local breeds. There was no set standard for its appearance.

Up until the early years of the 20th century, the Lincolnshire Buff was one of the mainstays of Lincolnshire farming. However, in 1860 William Cook developed the Buff Orpington which was extremely successful. The Lincolnshire Buff suffered and by 1920 it had almost disappeared. It was claimed that the Buff Orpington had been developed without the use of Lincolnshire blood; a statement that very few believed.

In 1987 the Lincolnshire Agricultural College began breed trials and through their efforts, and a few dedicated breeders, a breed standard was achieved. In the late 1990s the Lincolnshire Buff Society was formed. In 1997 the Poultry Club accepted the Lincolnshire as a standard breed.

Colour

The male has a rich orange back, neck and saddle with wings and tail which are copper and chestnut. The face, single comb, wattles and earlobes are red. The beak is fawn to white and the legs and five-toed feet are white.

25

Maran
Copper Headed Black

Native to
France

Now Found
Throughout the British Isles, Europe, North America and Canada

Description

In the 12th century the Duke of Anjou, later to become Henry II, married Eleanor of Aquitaine. Her dowry included lands in south-west France. British ships would call at La Rochelle, near Maran, where they would exchange the survivors of on-board cockfights for fresh food. These survivors would then breed with the local marsh hens. They are, very simplistically, thought to be the beginning of the Maran.

The breed, as we know it today, arrived in the British Isles in 1929 courtesy of Lord Greenway who brought eggs back from the Paris Exhibition. He started selectively breeding to standardise the colour. There are five standard colour variations in the British Isles and twelve in France. The Copper Headed Maran was first shown in the 1930s and, though not one of the famous five colours, it is an attractive and popular member of the Maran family.

The Maran is known for laying up to 200 beautiful chocolate-brown eggs each year as well as for its quality meat.

Colour

The Maran is black with a green sheen. The hens have rich copper coloured feathers on the head and neck and the cockerel has a lovely copper neck and saddle hackles. The beak is white or fawn, the eyes can be red or bright orange and the face, wattles, comb and earlobes are red. The legs and feet are white. The other colours are: Black; Dark Cuckoo; Golden Cuckoo and Silver Cuckoo.

26

Marsh Daisy
Wheaten

Native to
Lancashire

Now Found
A few worldwide

Protection category

Description

The Marsh Daisy was developed by John Wright in Marshside, Lancashire at the end of the 19th century. This work was continued by Charles Moor in the early years of the 20th century. Its name has two possible origins: the marshy area where it was developed, or the remarkable resemblance of its comb to the Marsh Daisy flower.

A good range of other poultry breeds went into the melting pot to produce this magnificent bird. The Black Hamburgh, the White Leghorn, the Malay and the Old English Game, with the Pit Game and the Sicilian Buttercup being added later.

This is a rare breed even in its homeland, the British Isles. It was at one time believed to be extinct but Ralph White, a past president of the Poultry Society, found a flock in the Williton area of Somerset in 1971. These remaining few are the ancestors of the present-day Marsh Daisy.

Originally there were five colours: Wheaten; Brown; Buff; Black and White. The Buff, Black and White are now thought to be extinct.

Colour

The male's hackles are rich gold, his back and wing bows are a deeper shade of gold and the body is a golden brown. The tail is beetle green-black. The face, rose-comb and eyes are red and the earlobes should be white but are usually pale yellow. The beak is fawn and the legs and feet are willow green with fawn toenails. There is also a brown Marsh Daisy.

27

Norfolk Grey

Native to
Norfolk

Now Found
In small numbers
throughout the British Isles

Description

The Norfolk Grey was developed at the beginning of the 20th century by Fred Myhill of Norwich and was first shown at the 1920 Dairy Show. This excellent dual-purpose breed was originally given the unfortunate name of Black Maria but this was soon replaced by Norfolk Grey. It is believed to be a cross between a Birchen Old English Game and a Partridge Wyandotte. The Norfolk Grey never achieved the fame it deserved, despite being a good all-round breed which lays about 230 eggs a year. In 1974 the breed was declared extinct.

The Reverend Andrew Bowden and his wife came riding to the rescue. They found a flock of four on a farm near Banbury which the owner was prepared to sell and the Norfolk Grey took the first steps to recovery. Once again, a breed was saved by an enthusiast.

The breed is still rare but, keeping my feathers crossed, it is safe for future generations. I am told that if you want to own a cockerel and have no neighbours, the Norfolk Grey is for you.

Colour

The male has a silver hackle, with black stripes. The back, saddle and wing feathers match. The remainder of the body is black. The face, single comb, wattles and earlobes are red, the eyes are dark and the legs and feet are black.

28

Old English Game

Black Breasted Silver Duckwing-Oxford

Native to
The British Isles

Now Found
Throughout Europe, America, Canada and Australia

Protection category

Description

The Old English Game has been strutting around the pens of the British Isles for over 200 years and is a descendant of the ancient fighting cocks known as Pit Game. During the Roman occupation a breed similar to the Old English Game was recorded. The breed has been around for over 2,000 years and during this time it has changed very little, if at all.

The Old English Game was bred for fighting before cock fighting became illegal in England in 1835. This attractive breed is now bred as a purely ornamental show bird. They still have an aggressive nature so, unless you are an extremely experienced poultry breeder, this is not the breed for you. If you put two males together they will fight to the death. The females can be equally aggressive and are very protective of their young.

Despite being noisy and refusing to be confined, they are still one of the most popular game birds. Choose any of dozens of colours and you cannot fail to be exhilarated by the dramatic appearance of this striking bird.

Colour

Both males and females have white necks and saddles with black breasts and thighs. Their backs, shoulders and wings are silver-white. The wings have a steel blue bar and the tail is black. The faces are mainly red with silver grey eyes and a white beak. The legs are also white. There are two varieties of Old English Game: The Carlisle and the Oxford. The Carlisle's back is horizontal to the ground and there are around twenty-seven colours. The Oxford, whose back is at 45° to the ground, has over thirty colours.

29

Old English Pheasant Fowl

Native to
The north of England

Now Found
In small numbers
throughout the British Isles

Protection category

Description

The Old English Pheasant Fowl has nothing to do with pheasants apart from its appearance. It was officially named in 1914, when the breed association was formed. Prior to this it had names given locally such as, Yorkshire Pheasant, Lancashire Silver Moonview, Manchester and Moss Pheasants as well as Silver, Golden or Black Pheasant Fowl.

Even before their official naming, the Old English Pheasant had been kept for over 100 years on farms in Yorkshire, Lancashire and what was then Cumberland and Westmorland but is now Cumbria. In all these years there has been little change in the Old English Pheasant Fowl's appearance.

This active and cold-hardy breed is known to roost in trees. It makes an ideal home or small farm fowl although catching one can be a challenge. Apparently, experience has shown that a landing net works the best. The eggs are white or slightly tinted and it is also an excellent table fowl. It is exceedingly rare, only 250 breeding pairs being known to exist.

Colour

The plumage is a rich bay and mahogany colour with darker lacing around the edges, a striped top and a lace breast. The hen is the same colour but with crescent shaped spangle markings. The rose-comb, face, wattles and eyes are red. The beak is fawn and the earlobes are white. The legs and feet are slate grey.

30

Orpington
Buff

Native to
The British Isles

Now Found
Throughout the British Isles, America, Canada and Australia

Protection category

Description

The Orpington was developed in a Kent village in the 19th century by William Cook, the son of a hostler who decided to work with chickens instead of horses. He became deeply involved in poultry as a journalist, advisor and lecturer. He made his small village of Orpington world famous. The breed is a cross between the Minorca, the Langshan and the Plymouth Rock.

The first Orpingtons were black and very similar in appearance to the Langshan. The other rich colour variations for this extremely attractive bird were developed later. This included the very popular Buff in 1894 in response to demand for buff-coloured birds.

This dual-purpose breed lays in the region of 180 brown eggs a year. The Orpington adapts to a small pen but free range is the better option; they have a tendency to overeat and plenty of exercise keeps them fit. Docile, affectionate and easily handled, they have a laid back approach to life which makes them an ideal pet.

Colour

The Buff have red faces, combs, wattles and earlobes; creamy white beaks and nails. They have orange or red eyes. The legs and feet are white. Other colours include: Blue, Black and White. There is also a Gold Laced variety and the rarely seen Jubilee.

31

Pekin
Partridge

Native to
America

Now Found
Throughout the British Isles
and on most continents

Description

The Pekin is a true bantam. One version of the Pekin's history says that they were liberated from the Emperor Xianfeng of China around 1860. Another story is that a number of these birds were given to Queen Victoria in the middle of the 19th century. These were crossed with other breeds for improvement, resulting in the Pekin bantam of today.

The Pekin are not the greatest egg layers with only in the region of 90 creamy-white coloured eggs a year but the bantam is small and could not sit on large quantities of eggs. The Pekin must be kept clean and dry: wet weather is a problem for this breed because mud will stick to the leg and feet feathers creating walking problems.

I have heard this delightful little bird described as a walking tea cosy. They are docile, love company and, if handled from a young age, will happily sit on your lap. An excellent choice for a first-time owner or children.

Colour

The male's head is dark red, shading through orange or red-gold down the hackles and becoming gradually lighter toward the shoulders. Each feather has a central black stripe. The breast, thighs, wings, tail and under-body are green black. The back, shoulders and wing bows are crimson. The hens have light straw or gold coloured hackles. Other colours are: Cuckoo; Mottled; Barred; Birchen; Columbian; Lavender; Silver; White; Buff and Red.

32

Plymouth Rock
Barred

Native to
America

Now Found
Throughout the British Isles, Europe, Canada and Australia

Description

This beautiful bird was developed in New England in the USA during the mid 19th century. Many breeders claim to have developed the Plymouth Rock but the honour has been given to John C Bennett who was also involved in popularising the breed. The breed is a result of crosses between the Dominique, Cochin and Black Java with a touch of Malay and Dorking. Prior to World War II it was the most popular breed in America.

The Barred variety was the first on the market followed by the Black and the White. The Buff did not appear until the end of the 19th century. When Plymouth Rock arrived in the British Isles in the early 1870s, breeders started immediately to improve their exhibition qualities.

This dual-purpose breed is docile, long lived and cold-hardy, making it an ideal bird for the small farmer or back garden owner. The Plymouth Rock lays in the region of 200 brown eggs a year.

Colour

The background colour is white with a blue tinge. They have bars of beetle green and a black tip to every feather. The face, earlobes, wattles and comb are red. The legs and beaks are yellow and the eyes are brown. The other colours: Black; Buff; Columbian; White; Silver Pencilled and the less common Partridge have the same facial colour.

33

Poland
White Crested Black

Native to
Poland

Now Found
Throughout the British Isles
and on most continents

Description

The Poland is an unusual and beautiful bird which is possibly the most popular of the crested breeds. It is also one of the oldest and has been known as a pure breed throughout Europe since the 16th century.

According to author Terry Beebe, 'The oldest reference found to date is a stone statue in the Vatican which bears a close resemblance to the crested fowl.' The Poland is commonly understood to have originated in Poland and Eastern Europe. There are unsubstantiated claims that it began in the Netherlands.

Originally kept for its production of white eggs, its main use now is ornamental. They are not suited to bad weather because they have an unusually thin skull which makes them susceptible to hypothermia. Their crest feathers are also liable to freeze. The feathers can also impair their vision causing them to be easily frightened, so many breeders tie the crest in a ribbon to avoid these problems.

Colour

A metallic black body and a brilliant white crest with black at the base. The face, small V-comb, wattles and eyes are red. The earlobes are white. The beak, legs and feet are blue or creamy brown and the soles of the feet are white. The male's crest is like an umbrella and the female's is like a powder puff. There are nine other colour variations.

34

Red Jungle Fowl

Native to
Asia

Now Found
Worldwide

Description

The Red Jungle Fowl is not a chicken but a breed of tropical pheasant. Charles Darwin determined that this bird, alongside the Grey Jungle Fowl, was the wild ancestor of the modern chicken. Archaeological evidence suggests that the Jungle Fowl was being domesticated and reared for food and eggs in India 5,200 years ago. Only a short time later it could be found in Europe.

I am told that during the mating season this flamboyant male will announce his presence with the traditional cock-a-doodle-doo. Nest building and incubation are left entirely to the female. That has a familiar ring to it. The female's less flamboyant colour is an effective camouflage when incubating the eggs.

This free-range fowl is not entirely flight-less. It can achieve sufficient height to spend the evening safe from predators in a tree or other roost.

Colour

The male has a golden orange head and hackle. The long, curved tail, which can be up to 23cm long, is a dark metallic green with a white tuft at the base. The under-body is black and the upper body is a riot of colour, rich orange, dark red and maroon. They have a bright red face and comb. The legs are grey. The hen is a dull golden brown.

35

Rhode Island Red

Native to
America

Now Found
Throughout the British Isles
and on most continents

Description

This is an amazingly successful breed and possibly one of the most recognised. It originated at the beginning of the 20th century in Little Compton on Rhode Island where it is the state bird. The breed was developed through a cross between the Asiatic Black-Red and a black-breasted Malay cock imported from the British Isles. This founding father is still on display at the Smithsonian Institution and a monument has been erected by the Rhode Island Red Club of America.

Hardy and disease resistant, they can survive better than most breeds on poor diet and housing although this is not advisable. The Rhode Island Red is an ideal breed if you want a small flock. Nine hens can lay up to six or seven eggs a day; depending on health and quality of life. This dual-purpose breed is used more for eggs than meat since a good female can lay between 250 and 300 brown eggs per year, although a good average is 170 to 200.

Colour

The plumage varies among deep shades of red. The male can have black or greenish-black tail feathers. The face, eyes, wattles, earlobes, beak and comb (which can be single or rose) are red. The beak is yellow.

36

Scots Dumpy
Black

Native to
Scotland

Now Found
In small numbers throughout the British Isles and Kenya

Protection category

Description

Scots Dumpys are also called Bakies, Crawlers, Creepers or Corlaighs. Similar birds were recorded over 1,000 years ago. The breed is now rare and I am told that the decline in numbers began with the Highland Clearances of the 18th and 19th centuries.

By the 1960s the Scots Dumpy was more or less extinct. Lady Violet Carnegie had taken a small flock to her estate in Kenya in 1902 and a number of these were returned to their homeland as breeding stock. They live on the estate of the Earl and Countess of Moray and these rare beasties are owned by dedicated breeders John and Sarah Burrows. It usually falls to small groups of dedicated breeders and breed club secretaries to keep rare and at-risk breeds from disappearing.

The Dumpy is best kept and happiest when free range. They have short legs (3.75 cm) so good level ground is an asset. They are excellent mothers, cold-hardy and a dual-purpose breed that has been known to lay up to 200 eggs a year.

Colour

Black with a green sheen. Their single comb, face and wattles are red and the eyes are amber. The beak, legs and feet are black or slate grey. The other colours are: Cuckoo; White; Brown; Silver and Gold. These other colours have white legs. The feet are white except in the Cuckoo whose feet are mottled.

37.

Scots Grey

Native to
Scotland

Now Found
Throughout the British Isles, America and Kenya

Protection category

Description

The Scots Grey is a very old breed whose history goes back to the 16th century, when they would have been a common sight on farms in Lanarkshire, Scotland.

This long-legged, erect-standing bird that struts around the yard like a toy soldier is an endangered breed. It is possible there are only 200 Scots Grey hens in the British Isles.

The Scots Grey is known for its hardiness and its ability to survive in harsh conditions. It has a great ability to forage and so requires plenty of space to roam.

Many are kept for purely ornamental reasons but the Scots Grey also provides quality-flavoured meat and large – considering the bird's size – cream-coloured, rich-tasting eggs.

Colour

The male is steel grey with black bars. The female is similar but with larger markings. The beak is white, sometimes with streaks and the legs and feet are white with black mottles. The face, earlobes and upright comb are red. The eyes are amber.

38

Sebright
Gold

Native to
The British Isles

Now Found
On most major continents

Protection category

Description

The Sebright is a true bantam. It was developed in the 19th century by Sir John Saunders Sebright and is one of the oldest bantam breeds. Charles Darwin was a great fan of Sir John, frequently writing of his expertise in the development and improvement of animal breeds.

The Rosecomb bantam was the basis for this striking breed which was crossed with British and Polish birds and a touch of Hamburgh and Nankin. Sir John eventually developed a laced bantam that would regularly breed true and the refinement continued until 1952 before finally achieving the standard we know today.

The male weighs in at 625g and the hens lay in the region of 125 small white eggs a year, so the breed is unlikely ever to achieve dual-purpose status. This purely ornamental breed is hardy, active, and friendly but skittish. This is not an easy bird for the novice to raise.

Colour

A rich, deep, golden bay with each feather edged in black, called lacing. The rose comb, face, wattles and earlobes are red or reddish purple. The eyes are black and the legs and feet are slate blue. The gold Sebright has a dark fawn beak. There is also a silver variety which sometimes has a dark blue beak.

39

Silkie
White

Native to
Asia

Now Found
Worldwide

Description

The Silkie is one of the oldest breeds of chicken. Its place of origin is most likely to be China, but Japan and India are also mentioned. The first known records of the Silkie came from Marco Polo in the 13th century who wrote an account of a chicken with feathers like fur. The breed arrived in Europe over 200 years ago via the Silk Road and maritime traders. These early poultry dealers sold them to the public as a cross between a rabbit and a chicken!

They lay 100 cream coloured eggs in a good year and do not lay at all during the summer months. The Silkie has black skin and bones due to the presence of a dark pigment called melanin. It is one of a few breeds of chicken to exhibit this. The grey-black meat is not generally acceptable to the European and American palate.

The Silkie requires little space and is calm, friendly, docile and trusting. This can, on occasions, be a problem when they are housed with more aggressive breeds because they can be bullied. This is the ideal first hen for children or a small garden.

Colour

The face, wattles and lumpy comb are mulberry coloured and the earlobes are turquoise blue. The legs and five-toed feet are grey. The beak is grey-blue and the eyes are black. The other colours are: Black; Blue; Gold; Partridge; Triple Laced Partridge; Triple Laced Silver Partridge. There is also a Bearded Silkie which is like a standard Silkie but with a beard and muffs.

40

Sumatra
Black

Native to
Sumatra

Now Found
Throughout the British Isles
and on most continents

Description

The Sumatra, usually called the Black Sumatra, is rare but the blue variety is even rarer. They come from Angers Point in Sumatra and were introduced into America in 1847. Fredrick R Eaton of Norwich introduced them into the British Isles in 1907.

Now kept as a purely ornamental breed, the Sumatra is exceedingly beautiful with the elegance of a peacock and a long, sweeping black tail. Its past was in cock fighting and the males frequently have the spurs to prove it. It is a pure breed, unchanged since its arrival in the western world almost 170 years ago.

The Sumatra lays in the region of 120 white eggs a year and because it is not weighty enough to be a meat bird it is kept purely for its stunning looks. It makes an ideal exhibition bird. They will not tolerate confinement and the more space, and the higher the trees in which to roost, the happier they will be.

Colour

The Sumatra is black all over with a rich green sheen. The face, pea comb, earlobes and beak are black. The eyes are dark brown and the legs and feet are black or olive green. The other colours are Blue and White.

41

Sussex
Speckled

Native to
Sussex

Now Found
Throughout the British Isles, Europe, America, Canada and America.

Description

The Sussex developed, strangely enough, in Sussex. It is a very old breed with the speckled variety believed to be the oldest although it is not mentioned in the Standards Book until 1865. The Sussex Club was formed in 1903. There is a school of thought that suggests the Old English Game bird was the starting point for the Sussex but, as in many breeding histories, proof is difficult to find.

This superb dual-purpose breed was once one of the main daily sources of meat and eggs for London. This venture centered on Tunbridge Wells and Eastbourne. The Light Sussex has been crossed with the Rhode Island Red, the Indian Game and the Leghorn to produce superb quality meat and still forms the basis for commercial production. The Sussex will lay in the region of 260 eggs a year.

The Red and Brown varieties are now quite rare, as is the Coronation which was bred to celebrate the coronation of King George VI.

The Sussex are good foragers and can adapt to most surroundings.

Colour

The head, neck and body are dark mahogany and each feather has a small white spot at the tip. The white spot is separated from the rest of the feather with a black bar. The tail feathers are black and brown with white tips. The face, comb, earlobes and wattles are red and the legs and feet are white. The Speckled, Buff and Red varieties have red eyes while the Light, Silver and White varieties have orange eyes.

42

Vorwerk

Native to
Germany

Now Found
In small numbers throughout the British Isles, Europe and possibly America

Description

The Vorwerk is a breed of chicken which was developed in Hamburg in 1900 by the German poultry breeder Oskar Vorwerk. The intention was to develop a medium-sized dual-purpose breed. It had to be gentle, hardy, economical, a good forager, ideal for a small farm and easy to look after. Ticking all these points, the Vorwerk is without doubt a 20th century success story. The breed was imported into the British Isles in the 1980s by Mrs Wallis of Arundel.

In 1966 William Vorwerk (no relation to Oskar) tried to buy Vorwerk chickens in America. When he was unable to find any he decided to breed his own. This was the beginning of the bantam Vorwerk which, apart from its size, is identical to the German breed.

These strikingly beautiful birds are good layers and will produce in the region of 170 cream-coloured eggs a year. Unfortunately, numbers are small and there is not even a breed club. They survive only because of a relatively small number of enthusiastic breeders and the Rare Poultry Society.

Colour

A deep buff coloured body with a velvety black head, hackle and tail. A red face, single comb and wattle with white earlobes, orange eyes and slate grey legs and feet.

43

Welsummer
Gold and Black Red

Native to
Holland

Now Found
Throughout the British Isles, America, Canada and Australia

Description

The Welsummer was developed in a small village called Welsum, on the banks of the river Ysel in Holland during the early years of the 20th century. It is a cross between the Cochin, the Wyandotte and the Leghorn and later development by a local farmer's son used the Rhode Island Red and the Barnevelder to help stabilise the breed. Prior to this improvement its appearance could have been described as a hotch-potch of colours and you would occasionally find five toes instead of the expected four. The Welsummer was imported into the British Isles in 1928.

This friendly breed is a good forager and will find most of its dietary requirements itself in a free-range lifestyle. They will lay in the region of 160 eggs a year which were described in *Fancy Fowl* magazine as 'a rich, deep, flower pot red-brown. Almost glowing'. A Welsummer will add colour and life to any farmyard or smallholding and you can see them every day looking at you from your cornflake packet.

Colour

The head and neck are a golden brown, the tail is black with a metallic green sheen and the breast is a black and mottled red. The single comb, face, wattles, earlobes and eyes are red, the beak is yellow or fawn and the feet and legs are yellow. The other colours are: Partridge, Silver Duckwing and Gold Duckwing.

44

Wyandotte
Blue Laced

Native to
America

Now Found
Throughout the British Isles and on most continents

Description

The Wyandotte originated in New York State and Wisconsin in America. They were named after the American Indian tribe, the Wyandot, who are possibly better known as the Huron. The breed's first column inches were in 1873.

The history of this comparatively modern breed is remarkably vague. It is a cross between the Cochin, the Silver Spangled Hamburgh and something else. The Silver version was the first colour variety but initially the result did not meet with the approval of the American Standards Committee. Further developments and refinements continued until 1883 when the Wyandotte was approved by the ASC.

The Wyandotte is often called the 'bird of curves' because of its wonderful shape. It is calm, docile and an excellent mother with the ability to adapt to free-range life or smaller garden runs. The Wyandotte is a superb dual-purpose breed, suitable for farm or show, and will lay up to 200 light to dark brown eggs a year.

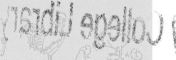

Colour

The feathers are reddish brown with blue lacing. The broad rose comb is red as are the earlobes and the eyes are dark red or orange. Other colours include: White; Blue; Buff; Red; Black; Barred; Silver Pencilled; Columbian; Partridge; Blue Partridge; Silver Laced; Gold Laced and Buff Laced. There are seventeen colours overall.

Chicken Talk

Cockerel – A male chicken

Hen – A female chicken

Pullet – A domestic hen of less than a year old

Chick – A young chicken

True Breeding – A breed in which the characteristics are consistently the same

Bantam – A small variety of chicken

True Bantam – A small variety of chicken with no larger counterpart

Dual-purpose – A chicken breed used to produce both meat and eggs

Autosex – A breed of chicken in which the appearance indicates the sex immediately after birth.

Appearance

Bars – Alternate stripes of light and dark across the feather

Comb – The fleshy crown on the top of a chicken's head. Usually red

Exchequer – A black and white checked pattern

Hackle – The long, slender, often glossy feathers on the neck of a bird

Laced – Having a coloured stripe or edging around the feather

Double-laced – Having a double stripe or edging around the feather

Muffs – The feathers which stick out from either side of the face, under the beak

Saddle – The back

Saddle Hackles – The long, narrow feathers on the saddle of a domestic fowl

Shank – The part of a chicken leg between the claw and the first joint

Spangle – A spot or shape on the end of a feather

Spurs – Sharp pointed protrusions on a cockerel's shank

Wattles – Two flaps of flesh that dangle under a chicken's chin. Usually red or purple

Wing Bows – The distinctively coloured feathers on the shoulder or bend of the wing

Toes – Most chickens have four toes although a few breeds have five with the extra toe pointing backwards.

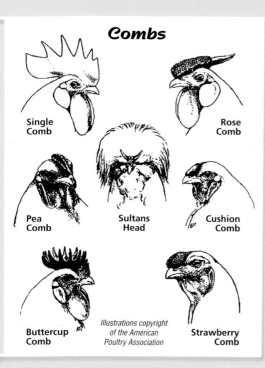

Combs

Single Comb

Rose Comb

Pea Comb

Sultans Head

Cushion Comb

Buttercup Comb

Strawberry Comb

Illustrations copyright of the American Poultry Association

RBST
Rare Breeds Survival Trust

The Watchlist covers sheep, cattle, goats, poultry & ponies.

A breed whose numbers of registered breeding females are estimated by the Rare Breeds Survival Trust to be below the Category 6 'Mainstream' threshold will be accepted into the appropriate Watchlist category.

In this book I have highlighted the first five categories.

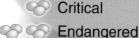

Critical

Endangered

Vulnerable

At Risk

Minority

Further information: www.rbst.org.uk

Also in the 'Know Your' series...

Know Your Sheep
Forty-one breeds of sheep which can be seen on Britain's farms today.

Know Your Tractors
Forty-one tractors working today including classics and modern machines.

Know Your Cattle
Forty-four breeds of cattle. Some you will recognise and some are rare.

Know More Sheep
A further forty-four sheep breeds including popular mules and a few unfamiliar faces.

Know Your Horses
Forty-three breeds living around Britain today and a few rare ponies.

Know Your Combines
Forty-three combines you are likely to see in the summertime fields.

Know Your Pigs
Twenty-eight breeds of pigs in every shape and size. Some you will know, and some you will have to look hard to spot.

Know Your Trucks
Forty-four example of trucks that you are likely to see driving Britain's roads today.